PERSONALIZED TECH

PERSONALIZED TECH

NOLAN BLACKWOOD

CONTENTS

1 Introduction to E-Ink Technology and Wearable Devi   1

2 Current Applications of E-Ink and Wearables   5

3 Benefits and Advantages of E-Ink and Wearables   9

4 Revolutionizing Personal and Professional Life   13

5 Future Developments and Innovations   17

6 Challenges and Limitations   21

7 Conclusion and Final Thoughts   25

Copyright © 2024 by Nolan Blackwood
All rights reserved. No part of this book may be reproduced in any manner whatsoever without written permission except in the case of brief quotations embodied in critical articles and reviews.
First Printing, 2024

CHAPTER 1

Introduction to E-Ink Technology and Wearable Devi

In the field of low-power and mobile devices, because electronic ink is reflected light-based manufacturing, the ambient light will help increase the visual effect. Its reflective nature is a direct view feature that adds natural viewing and a luxurious way to improve product quality. Most of all, electronic ink has a kind of stable property that usually only needs to be replaced when generating changes. The reduction color and true multi-color technology of electronic ink can also make it more youthful. There are even opportunities to produce electronic ink for the display of digital information that can be unique and changes in appearance. With the continuous development of applications, the application and manufacturing of electronic ink will be further developed.

Electronic ink (e-ink) has become increasingly popular recently with an amazing user experience. It has excellent bi-stable and light reflection and can be easily carried around. It is a low energy consumption device. The university application and wearable market is rapidly expanding. Because of its low-power and outdoor read ability, as well as the lack of screen flickering, it is a gentle display of the potential and advantages of e-ink. This article provides and discusses electronic ink technology applications in detail. In recent years, to improve the lighting requirements and the emergence of color e-ink technology, the university application has been increasing, and more manufacturers are investing in research and development. The color enhancement of e-ink is compatible with soft solutions. Today, electronic ink technology is not only applied to e-books; in addition to smartwatch sports watch wearable devices and intelligent personal server panels, there are more diversified applications that have been launched in the market. In the personal field of view and cognitive experience, it can significantly influence the development and expansion of wearable device applications. This paper provides a comprehensive solution that utilizes e-ink technology for wearable devices to provide personalized technical ingenuity.

1.1. Definition and Basics of E-Ink Technology

The first flexible E-Ink display was a 4-inch diagonal watch, having a resolution of 80 pixels × 28 pixels with 130 micrometers on a side. The watch could display any four digits in veritable style according to the time (e.g. 1:00, 12:59). Since this prototype, there have been a variety of improvements made in order to personalize the qualities of E-Ink. While displaying 4000 elements per square inch and keeping resolution problems under control is crucial for a visually pleasant experience, equally important is the efficiency extending the lifespan of a battery. The E-Ink displays are made to be rather efficient, presenting color accelerations and minimal required energy. This, again, saves energy.

E-Ink is a display technology, invented at MIT, that simulates the appearance of regular ink on paper. As a bi-stable medium, each semi-transparent microcapsule of ink contains negatively charged black and positively charged white particles with opposite charges that are rotated by an electric field. These particles are held in a droplet with a clear positive and black anode at the bottom, which gives the display medium its bi-stable property. Once a particle is rotated into visibility, it stays that way unless manipulated by another charge. Thus, the E-Ink display consumes very little energy - power is only required when changing the display. This forms the basis of flexible, low-energy displays that can be improved and personalized in the future.

1.2. Evolution of Wearable Devices

In the case of E-ink, high reflectivity and an absence of backlight reduce power consumption, allowing for information to remain displayed without connection to a fusion system. In the case of FDUs, the development of a low (less than 2V) operational voltage and simple control and data interface solution (implying "direct addressing") now allow the production of high resolution, large-size monochrome- or multicolor-capable field-sequential display types for the wearables market. The FDU may be realized by printing or a photophysical approach on a mechanically flexible plastic foil or an ultrathin glass foil. The actual availability of such low lightweight FDU panel technology supports a "see-through" technique for the reactor chamber windows used in industrial waste gas monitoring applications. The lightweight, robust and thin form factor of the flexible E-ink display is also very attractive for use in wearable devices and clothes-on-the-move with high reflectance-adopted near "paper-look" user friendliness.

The current scenario points to interactive/ubiquitous wearable devices and clothes with the provision for controlling electronic devices, writable flexible display units (FDUs), and personalized content. This is all being made possible by the growing use of flexible electronics, mechanical flexibility, and the development of form factors of displays whose physical property imparts flexibility and lightweight functionality. However, much more compact, lightweight, robust and simple "SoC" devices have recently been developed for realizing observer independent operation, used for daily life applications. The "post-PC" and "post-smartphone" concepts are particularly important in this respect. The history and present status of electronic displays are briefly reviewed before discussing wearable displays. These cover E-ink, one of the most successful displays and the ultimate thin and lightweight display of a flexible FDU with near E-ink contrast.

CHAPTER 2

Current Applications of E-Ink and Wearables

Adjacent to the display of the latest status of items in retail and warehouse scenarios, wearable technology based on e-ink is more suitable for adoption in significant medical applications such as smart prescription glasses and smart contact lenses for real-time message feedback. The vision for fully foldable and bendable smartphones is not far from reality. Ingredients such as organic semiconductors, organic photodetectors, tactile sensors, wireless power and communication modules, and different kinds of flexible technologies are ready for this new wave of technology. Smart driving licenses can be made with e-ink and other smart card technologies for identity authentication and security reasons. Finally, e-ink wallets for secure payments in conjunction with e-ink credit cards also have good potential.

The development of wearable technology has important implications for industries from IT to healthcare. Most current devices are too large, impractical, and uncomfortable to wear, but the development of e-ink will allow wearable technology to be smaller, sleeker, and more comfortable to use. Applications of e-paper and e-ink today include digital signage, electronic shelf labels, e-books, indoor and outdoor advertising, and electrophoretic displays in mobile handsets. Consumer goods bearing simple displays can be updated for evergreen appeal. Flexible price tags using this advanced technology will remove human errors involving price labeling, reduce labor costs, and provide more accurate pricing across the wide range of consumer stores around the world. Industry applications include area identification signs, portable environmental sensors, ultra-low power ID tags, device labels, and wearable electronic patches.

2.1. E-Readers and E-Paper Displays

There is also research in the color e-paper and large-area e-paper fields. Non-black and white e-paper displays are expected to meet smartwatch usage in the future. Because of the increasing market demand and product iteration speed, e-view paper shipments are expected to maintain their growth momentum in 2020. Following the future trend as tablets do to touch laptops, will smartwatches also affect the sales of overall wearable devices? Will other e-paper-form-factor wearable devices benefit more in the future? As thin and light e-paper products become more widely used, can shipments keep growing? The market trend of wearable devices is still one of the main influences determining the growth or reduction of e-paper shipments.

Smartwatches are another market for e-paper and have seen explosive growth. These devices use black and white displays because of their low power consumption and excellent visibility in direct sunlight, and there is currently no better alternative. Developers are expected to increase wearable OS for scene-specific applications according to user behavior and habits, making hardware and software development more focused and more specialized. This demands not only a unique and precise stress-sensing module but also the development of real-time, large multi-feature identification and posture identification of the algorithm by better machine learning techniques, and increased understanding of the issue of user privacy.

E-Ink Holdings is currently working with seven universities in Taiwan to develop e-readers with colored e-ink screens for e-learning. Their goal is to popularize the use of light, low-power, and powerful e-readers while decreasing the weight of textbooks. The project aims to create learning books that combine print and e-paper and to set up an analysis system to measure user reading habits, allowing for further improvements in interactive techniques.

Despite a slowdown in e-reader sales, the technology is still relevant for digital education. Textbooks with interactive web-based curriculum material have become the standard for K-12 learning, but the majority of students still use heavy printed books. The enhancement of this e-paper technology will make these digital learning materials lighter and more appealing.

Devices such as the Kindle and Nook brought attention to e-ink and e-paper displays when the e-reader market took off. With better contrast and similar viewing quality to the printed page, displays with an e-ink screen are easier on the eyes and offer a distraction-free reading experience. E-ink displays offer low power consumption and good sunlight readability, but the trade-off is slow page refresh rates. This slow page-turning speed may not be a significant issue, considering that novels are not typically read at the pace of a comic book.

2.2. Fitness Trackers and Smartwatches

But because e-ink and also transreflective TFTs have practical advantages, such as sunlight visibility, reflective nature that empties batteries in use only if the image changes, and their performance in reading a large quantity of small- and micro-pixels, these should be developed with a greater variety of displays and functionalities. The main advantage of transmissive displays is the wide availability of color because red, green, and blue LEDs are monochromatic, so the pixel is colorable when they are crossed. There is still some transistor paradox on the back or top-emitting OLEDs, and due to the small size and relative simplicity of the technology, a project that would include 72 colors and 72 rows would not add much complexity or cost to the current smartwatches, fitness bands, and other available devices.

Fitness trackers, specifically wearables like the original Fitbit and the Jawbone devices, pushed the adoption of e-ink technology by the world's largest consumer electronics manufacturer, Apple. Initially, the use of e-ink was not persistent in the device as it is in dedicated e-readers because, in order to receive notifications, read messages, change the status of the music, and check the time, the display had to be activated through taps, squeezing, sometimes pushing buttons, or relatively non-practical gestures, such as quickly rotating the wrist. Although useful as a means of enabling activity and sleep tracking, these devices share a common feature that limits them to these activities. The display size is suboptimal for interaction with notifications, limiting them to the aforementioned gestures, unlike smartwatches that use color or round OLED displays.

CHAPTER 3

Benefits and Advantages of E-Ink and Wearables

There has been a push to migrate viewers from glossy screen displays to the more reader-friendly E-Ink. Technology companies see E-Ink as a viable alternative to glossy screen displays because of their long battery life on the order of days or weeks, use in direct sunlight, and ideal applications for e-commerce. E-Ink displays are designed to modify the display of text when interactively touched by a user, read the current display of the multifunctional user device, browse the network to receive the new text from the network, and wait for a predetermined time period. We have developed a system that has a new design for using the benefits described above for disposable, unobtrusive, and low-cost e-commerce devices in wearable use.

E-Ink and wearable technology (WT) are quickly becoming the future for many tech enthusiasts. E-Ink devices are popular with major technology companies for their long battery life, high contrast, and flexibility in design because they are able to be bent. With the increase in e-commerce, today's workforce is always on the go and even on call. Technology advances have allowed customers to access the internet from almost anywhere and are no longer limited to a desktop computer for this ability. As such, mobile devices such as smartphones and tablets are used to read websites and e-books, update online social media profiles, and purchase from a variety of e-commerce businesses, while taking little room in a user's pockets, bags, or briefcases.

3.1. Energy Efficiency and Battery Life

Lowering the overall power consumption in the design and optimization of the hardware and operating circuitry, especially in memory components, transistors, and integrated circuits, reduces the clock rate at the expense of lowered power consumption and are energy-efficient ways of increasing the time in which the device could intermittently be turned completely off. Additionally, architectural designs for energy efficiency in the system's memory, display, and PCB board could contribute significantly towards power optimization. Since both the e-ink and wearable devices are intended for long periods of use, the batteries are likely to have degraded significantly by the end of their life cycles. During the integrated circuit layout and design phase, the circuitry and voltage levels should be designed so that the device could still operate with the batteries Vmin of up to 2.7V (using a lithium polymer battery common in mobile applications of similar size). As such, thin-film transistors in the target e-ink displays should have a low threshold voltage to enable operation at significantly lower power supply voltages without needing any extra peripherals such as regulators.

Energy efficiency is a longstanding issue in the design of consumer electronics in general. However, with e-ink and wearable devices, energy efficiency is even more critical due to the prolonged period of use of these devices between recharges. As such, in the case of our research, while energy efficiency is a key factor of interest in the design of both a personalized e-ink and a wearable device, energy optimization involves a lot more than the traditionally discussed power-saving techniques. Power saving mechanisms used in previous applications are largely based on powering down or reducing the functioning of components which are not needed. However, due to the continuously updating and interactive nature of the proposed personal e-ink and wearable electronic devices, these traditional techniques are not applicable in their stagnant state which is the main source of power consumption.

3.2. Customization and Personalization

One of the truly great things about technology - both hardware and software - has been the personalization and customization it allows. No longer must a user learn to live with the way a product is set far from the factory, or choose from a few preselected options. In fact, many high-tech items, like smartphones and tablets, are actually downloaded slates that users then fill with the software titles and settings they desire. Because the technology world is ever-evolving, ever-sprouting, nascent areas like digital wearables and flexible, wearable displays don't come with many consumer expectations. That's about to change, thanks to some well-known companies diving headfirst into the field and setting standards. Locally, Electrophoretic Display Holdings is hard at work designing and perfecting a line of E-Ink powered PDA-sized digital reading devices that can fold up and fasten to a belt buckle. The appeal of this clever, mobile reader is GIF-netting: A whole new meaning for the E-Ink powered map application, for starters. Soon to supersede maps, even a low-res PDA display would have a hard time competing with a line by line, E-Ink direction inbox, which is like having one hand on a steering wheel and the other ready with a magic marker to highlight the newspaper on all the public streets.

CHAPTER 4

Revolutionizing Personal and Professional Life

eInk technology has been widely used in e-book readers and is also being applied in commercial and education-related electronic paper products. Taiwanese manufacturers are transforming notebook computers into electronic notebook products, which has prompted them to gain popularity and become the next wave globally. The magic of eInk technology enables manufacturers to create products with distinctive features and professionalism.

Driven by increased mobility, the information technology pattern has shifted to small screens that can be carried around, are energy-saving, and have rapid human-computer interaction response times. Mobile device development, especially in e-books, cell phones, and other terminal applications, has become faster and more widespread. The eInk technology adds new vitality to mobile devices and enables them to provide a paper and ink reading experience. It is expected that eInk mobile devices will have a bright future in electronic newspapers, electronic reading, and other display devices. Since the 1990s, electronic newspapers, electronic textbooks, electronic reading, and other applications have been popular in the North American, European, and Japanese markets. Even with the rapid development of technologies such as color readability, ink interface, and touch screen, eInk technology has achieved huge market penetration due to its environmental protection and energy-saving concept.

4.1. Impact on Health and Wellness

Smartwatches and fitness tracker wearables can impact the user's health in a somewhat different manner. These products are often marketed using sports challenges and training objectives, but smartwatches can also be used to store personal health information such as blood type, allergies, medical history, and medication records, making it possible to receive quick medical care upon notification by pressing a button on the device. The fitness tracker in a smartwatch or wearable can support its user's health by measuring steps taken, calories burned, and even the heart rate. Would you like to broadcast this information to data streams that supply that health data to the cloud?

Let's start with practical use, as the health benefits of a reflective display might be considered by some to be more tangible than the broader advantages it presents. As paper is by nature less strenuous to the eye than an electronic display of photons, using an e-ink reader for e-reading can strain the eye less than an LED, LCD, or OLED-based device. An e-ink reader's screen refresh rate is slower than those of the previously mentioned display techs, but more and more e-ink readers are available with refresh rates that allow them to be used not just for reading books and monochrome newspapers, but also for reading color comics. This takes away the need to decide whether to use a paper or pixellated display for those printed products.

4.2. Enhancing Productivity and Efficiency

Given that technology use is time and task dependent, wearable tech will opt to cross paths with the user during the most optimal times and parameters, defined by what is currently understood in scientific research about individual cognitive and physiological work rhythms. Recent research may uncover additional patterns that could be applied to this new technology to increase its effectiveness. That way technology can move from being more of a general aid to a more discriminating arbiter of those periods in which it will be of optimal assistance.

Minimizing the time necessary to interact with devices is not the only way to bolster personal productivity; figuring out exactly when, and how, to interact can improve the outcome. Studies have confirmed that alertness, energy, and even efficiency at completing various tasks on the job follow distinct patterns that change over the course of a work day. Some of us are "morning people," raring to go when the rest of the world is still rubbing the sleep out of its collective eyes. Others tend to hit their stride in the mid-afternoon and wind down after we've eaten our evening meal. This means that the best time to perform different sorts of tasks - whether they are decision-making, analytical thinking, or socializing - can vary from one person to the next.

CHAPTER 5

Future Developments and Innovations

The conversation ability of high-tech speech recognition has begun to replace many interaction devices, including Arduino and Raspberry Pi. For example, a magic episode has been developed by combining voice chips with e-paper technology. Users can communicate with the speaker through human dialogue and switch to a method of manual manipulation at any time using predefined trigger words or phrases. The LED array display allows users to see the visual description of the content transmitted by the current device. To do this, the reading display installed on the input device can send the data to the microcontroller, and the speech synthesis designed with the speech synthesis software voices the syntax and word used in the speech content. The questions and answers in human-machine communication are processed by the voice chip.

In the wake of increasingly popular devices such as smartphones, which can stream video, make calls, and allow users to surf the internet, the electronic paper industry is evolving in the direction of the high-end market. Such new, personalized technologies can be products such as embedded AI audio speakers, 5G, and eSIM. The rise of personal technology is expected to drive the growth of TFT capacity. Despite the increasing trend and slowing market expansion speed of personalized devices, the concept of personalized technology has been consistently prevalent in various electronic fields. However, the psychological stimulation based on futuristic concepts that dominate in personalized devices is relatively unbreakable. Consequently, in this study, the future trends of personalized technology are explored by providing current TFT processing methods, physical methods, and electronic methods, to reveal the significance and the potential market of this research area.

5.1. Flexible E-Ink Displays

Another novel method of obtaining flexible E-ink displays is through the application of E-ink active matrix driven flexEnable technology. This uses a backplane integrated into a sheet of plastic and serves as a top layer for monochrome panels. These top layers can be chosen from a range of materials and properties, enabling flexibility as an option and can be driven by the backplane, enabling video or other fast update capabilities. The white reflective part of the E-ink display is created on top of the backplane, which can be either a soft plastic or glass, meaning that the entire display is suitable for flexible e-paper or e-notebook applications.

Bendable plastic displays are the main perceived advantage of E-ink technology over traditional liquid crystal displays. So far, however, developing flexible E-ink displays suitable for high volume manufacturing remains a challenge. To make an E-ink display flexible, greater flexibility must be integrated into the cathode, either by adding flexible organic electronics to the cathode itself or by adding a thin film transistor to the backplane. The addition of this transistor affects E-ink's core technology, so either the cathode must be adapted to be manufacturable in sheets that can later be combined with a top layer, or the top layer must be made flexible.

5.2. Integration with Internet of Things (IoT)

If we talk about the Internet of Things (IoT), one of the very significant goals consists of increasing computational power and connectivity while reducing the dimensions and power consumption of the physical devices. A natural emergence in this concept is related to the use of portable and wearable displays. The traditional displays, mainly used in smart devices, are made by OLED or LED technologies, thus resulting in high energy consumption by the display. The solution is represented by the e-Paper displays, which can maintain the displayed information even without the energy supply. Our study proposes a new family of PCB-mounted e-Paper display drivers, targeted to low-cost and simple IoT products. Thanks to the low power of the e-Paper display, the risk for malfunction in the IoT nodes created is significantly reduced.

E-paper/ink can bring a unique sense to integrate with the Internet of Things, not just as a display. It has characteristics such as a paper-like surface, thinness, low power consumption, and flexibility. These characteristics reduce the difficulties for e-paper/ink adoption in IoT sensors. Furthermore, e-paper/ink can employ IoT to sense keywords from documentation, search, and present relevant entered information to the user. When e-paper/ink is the display of a wearable device and installed on clothing, most e-paper/ink technology is loosely curved. Therefore, e-paper/ink can be e-cloth in terms of the Internet of Textile and IoT technology. In this kind of work, studies of image recognition by utilizing machine learning and image assessment by employing medical research are presented, which can represent these capabilities of e-cloth and be applied in action.

CHAPTER 6

Challenges and Limitations

There are notable challenges and limitations in current E-ink technology, including relatively slow display speed and limitations of device design and material. The high rigidity of inorganic materials requires extra protection layers of plastic or glass, which limits the flexibility of E-ink. Flexible polymer-based active matrix backplane technology, such as flexible single-crystal silicon (FSCS) backplane, is developed to address the flexibility issue but with limited commercial products. High drive voltage is another significant limitation for E-ink, especially for wearable and Internet of Things (IoTs) applications, which require low energy consumption and small device size. Additional breakthroughs are required to develop a new generation of E-ink technology that features low power, small footprint, thin, and easy to mass produce for various current and future applications.

E-ink technology has come a long way and with the increasing demands for e-Paper technology, it will improve and evolve further. Presently, some e-Papers are not able to support apps that require continuous data exchange. However, the issues with color e-Paper, effective integration with smartphones, and the overall electronic stability, weight, power requirements, costs, and other key technical specifications will improve in the near future. Nowadays, the assembly of the electronic clothing components requires reliable assembly and attachment methods, whether for body sensors, light-emitting tiles, or flexible e-Paper displays. Thus, the electronic clothing modules can be exchanged if needed while preserving the clothing design.

6.1. Technical Constraints and Performance Issues

In line with our initial promise to the reader, we provide in this section, within the theoretical and experimental performance data, a description of E-Ink display which will enable the reader to replicate the experiments described in detail. So that there will be no misunderstanding regarding the difficulty or otherwise, if the optional route is included, an analysis of fill-density and thickness versus voltage respiration with optical drive and performance measurements is included. For high-quality imaging, there is therefore no current operation path to the id by the three together sets of data. Increased pixel or pixel geystation reduces reflector speed and decrease reflectance contrast and brightness with the final restore procedure. Infinite cooking typically begins at very low disuse home voltage.

The hierarchical organization of the E-Ink displays on a page basis, primarily due to the easy availability of address voltage programming, has not been sufficiently used until now. In fact, from a design of technology performance point of view, the best results should be obtained by individually programming the address voltage of each pixel. This approach is difficult and, in some cases, nearly impractical, so that custom imperfections determination cannot be avoided. Most image devices need full programming even if not all pixels change state. These results have established a series of constraints which must be met in order to obtain accurate data and for seamless inclusion of E-Ink displays in image technology.

6.2. Privacy and Data Security Concerns

As a matter of general strategy, companies should also prefer a transactive IoT business model (over a surveillance capitalism business model) for the products in this category. If a business generates a significant part of its customer service successes on the back of well-designed software, then there should be things to sell apart from the actual tech gadgets. Public knowledge and the automatic disempowerment that comes from it are a guarantee of inherently strong open-source development ecosystems.

From the very beginning, it was clear that there would be privacy and data security concerns related to being able to accurately and remotely monitor the physiological data of users. If this kind of product is going to be marketable, then first and foremost, privacy must not be violated. Today, companies should already know that no amount of electronically collected data should ever leave a customer's device. While any balance-focused tech business absolutely needs and wants detailed, personal usage statistics to improve their products from firmware to ecosystems, uplink data to the balance should be anonymized or not used at all. The latter reason is very much safety-related since any kind of unauthorized access also includes the possibility of unauthorized manipulation.

CHAPTER 7

Conclusion and Final Thoughts

The new generations of flexible photovoltaics are likely to provide flexible energy sources that could be curved or pleated to the form of a device before covering it with the encapsulation. It is also expected that future chip manufacturing will provide fully flexible and high-density integrated circuits ready to be deployed, allowing the user to concentrate on how to knit the components together. Finally, we believe that light-street fashion could intercept a new market for e-paper, where personalized renders of user-generated artwork and photography dominate over general readers of e-books and e-papers, leading to a new type of initiative emerging to deliver similar content. This could revitalize user demand for small-scale local publishing.

In the future, the display will be part of the culture of the maker and help create unique identities for the owners of the devices. We note that some very odd, quirky, and extremely interesting opportunities are opened by the combination of flexible displays and fully flexible systems. Some ideas are given in the supplementary materials, but many more await only creative thoughts to realize them.

This article has presented novel advances allowing the manufacturing and use of fully personalized mobile e-paper devices. Prototypes demonstrated combinations of various customization patterns and integration with different rigid and flexible electronics components. We conclude that these advances provide exciting new opportunities and a shift in perspective for e-paper displays. We also note obstacles still to overcome, and areas that present interesting future research challenges.

Milton Keynes UK
Ingram Content Group UK Ltd.
UKHW031021011224
451693UK00004B/557